INVISIBLE INK

Written by
Read With You Science Academy

Read With You

The characters and events portrayed in this book are fictitious. Any similarity to real persons, living or dead, is coincidental and not intended by the author.

Published by Read With You Publishing
Designed by Read With You Science Academy
Read With You and associated logos are trademarks and/or registered trademarks of Read With You L.L.C.

ISBN-13: 978-1-64616-112-6
ISBN-10: 1-64616-112-2

Third Edition April 2021
Printed in the United States of America.

Welcome to the

READ WITH YOU SCIENCE ACADEMY!

Today we will do an experiment called

INVISIBLE INK

Before we begin, let's review some important words:

acid:
A chemical with a sour taste that forms a salt when mixed with a base.

base:
A chemical that reacts with an acid to form a salt.

to react:
To change after coming into contact with another substance.

QUESTION:

What I want to know.

Can I make invisible ink?

- [ ] Yes, I can make an invisible ink.
- [ ] No, I cannot make an invisible ink.

Remember:
Be a careful scientist! Clean up any water that spills right away!

HYPOTHESIS:

What I think will happen and why.

I **can** make invisible ink because...

I **cannot** make invisible ink because...

I **can** make invisible ink by using baking soda because:	I **cannot** make invisible ink by using baking soda because:
Baking soda is a kind of salt.	Baking soda is used for baking, not writing.
Baking soda is a base that will react with grape juice.	Baking soda will not react with grape juice.

What do you think?
Circle your hypothesis!

EXPERIMENT:
How we will test our hypothesis.

THINGS WE NEED

- Baking soda
- Paper
- Water
- Light bulb (heat source)
- Paintbrush or swab
- Measuring cup
- Purple grape juice
- Tub or tray

baking soda

paper

water

light bulb

paintbrush

measuring cup

swab

grape Juice

tray

PROCEDURE:

The steps to complete the experiment.

Step 1A

- Mix 1 teaspoon of water and 1 teaspoon of **baking soda** in a glass bowl.
- Put the glass bowl in a larger bowl or on a tray.
- Use a paper towel to clean up any spills.

Step 2A

- The baking soda solution is your "ink." Use a cotton swab or paintbrush to write a message on white paper.

Step 3A

- Your paper will look blank.
- Allow the ink to dry.

Step 4A

- Paint over the paper with purple grape juice. The secret message should appear.

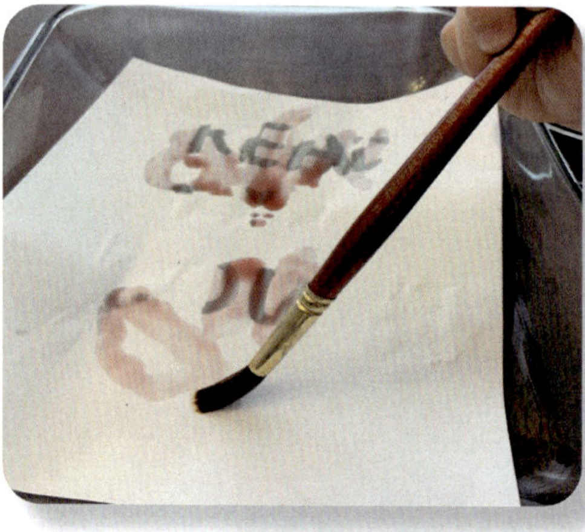

READ
with
you

Write down what you see

In your science journal,

write what you **observed!**

- What was your **hypothesis**? (Look on page 5 of this book!)
- Was your hypothesis correct?
- How quickly did your message appear?

Let's do it again!

Can a heat source be used to read the secret message? Let's do an experiment to find out.

Don't forget to ask an adult for help with this experiment.

QUESTION:

What I want to know.

Will heat make the invisible ink become visible?

HYPOTHESIS:

What I think will happen and why.

I **think** holding the paper to a heat source will allow me to read the message because…

I **don't think** holding the paper to a heat source will allow me to read the message because...

What do you think?

Write your new hypothesis in your science journal.

PROCEDURE:

The steps to complete the experiment.

Step 1B

- The baking soda solution is your "ink." Use a cotton swab or paintbrush to write a message on white paper.

Step 2B

- Allow the ink to dry.

Step 3B

- Hold the paper up to a heat source like a light bulb. Do not use a halogen bulb.
- Observe what happened.
- Don't forget to ask an adult for help with this experiment.

Hello

Write down what you see

In your science journal,

write your results!

- What was your hypothesis? (Look on page 11 of this book!)
- Was your hypothesis correct?
- Were you able to read the message?

Write the message in the box below. Which method worked better, the grape juice or heating the paper?

Why did this work?

- **Cellulose** is a substance that is the main part of the cell walls of plants. It is used to make various products, such as paper.
- When you wrote a secret message using the baking soda solution, it made a small change to the cellulose fibers in the paper.
- When the paper is heated, the baking soda causes the writing on the paper to turn brown. The grape juice acts as a **pH indicator** (a chemical compound added to a solution to determine if it is an acid or a base) that changes color when it **reacts** with the baking soda.

SCIENCE

- Almost every liquid is either an **acid** or a **base**. You can use acids and bases in your everyday life, especially in the kitchen.

- **Baking soda** is a base that makes bubbles when combined with liquid and heat.
    - These bubbles make baked goods like cakes, muffins, cupcakes, and cookies rise.
- Acids are used to break things down. Vinegar is an acid that can be used to clean windows and other surfaces because it breaks down grease.

GLOSSARY

Acid: A chemical with a sour taste that forms a salt when mixed with a base.

Base: A chemical that reacts with an acid to form a salt.

Baking soda: A kind of salt that is used to make baked foods light and fluffy; also called sodium bicarbonate.

Cellulose: A substance that is the main part of the cell walls of plants.

To dilute: To make (a liquid) thinner or less strong by adding water or another liquid.

Hypothesis: An educated guess; what we think will happen and why.

To observe: To watch something carefully.

pH indicator: A chemical compound added to a solution to determine if it is an acid or a base.

Procedure: The steps to complete the experiment.

Question: What we want to know.

To react: To change after coming into contact with another substance.

NEW IDEAS

How could we change this experiment?

- Is there anything else that could be used to create the invisible ink?
- What would happen if we used apple juice instead of grape juice?

Can you think of any ideas to improve the experiment?

- The baking soda could be **diluted**.

JOURNAL

Made in United States
Orlando, FL
07 November 2026

90337518R00017